ENTERING THE COMPUTER ERA

By

Karen J. Pleban

INTRODUCTION

A view of a retired computer programmer's thoughts presents the basic logic and rapid growth of computers with some of the adversities that has developed with the growth of computers.

The information I provide will touch on the evolutions of computers starting with the beginning in the Ice Age times into the millennium years with consequences. The march of computer competition in many forms started out with a very simple logical method of calculations.

Man's ideas on calculations have continued to progress beyond one's imagination into today's millennium era with the Personal Computer commonly referred to as the PC, which affects all of us one way or another in our every day lives. Certain negative consequences developed during this phenomenon of computers and continues on to date.

The uniqueness is that each progression of the computer from the beginning had originated from humans with visions creating reality of tomorrows, proceeding forward with the uncertainty of the results.

The contents and the research in this book are combined from my observance, experiences and a small collection of information acquired purely for my personal interest in the subject. (See

resources)

Table of Contents

List of Illustrations   **Page**

Acknowledgment

The author wishes to express sincere appreciation to her husband, Ronald Pleban for his encouragement, assistance and introducing the idea of entering the computer field.

Back in the early sixties when we were dating, Ron was enrolled in an Electronics Certification Program. We spent most of our dates studying the logics of electronics. As a result, soon after our marriage, we started a radio and television repair shop. In order to assist with the business, Ron taught me the electronics color code, reading transistors, how to test television tubes and solder wires. Ron's inspirations planted the roots to my success.

In 1973, Ron suggested that I enroll into full time course in a specialized computer school. The very thought of my being able to understand the computer logic was intimidating let alone entering into a predominately male field of expertise and two little boys at home, but Ron encouraged me to proceed, hence I acquired my Certificate of Computer Systems/Analyst programming. While in school, I served my apprenticeship as a keypunch supervisor. After graduating from Computer School, the first job assignment I acquired was to convert the entire manual accounting system in a small engineering office into a computerized system. This was with an IBM 5100 which used cassette tapes for storage. This proved to be a challenge for my first programming/analyst job. Following that job, I was a systems analyst on a larger system with the COBOL and Neat/3 languages in manufacturing environments. I became strong in software, however with the evolution of the Personal Computers, my husband Ron further pushed me to the understanding the hardware portion of the computers. As a result, I have attended numerous classes and received an array of certificates to include the A+ certification.

THE BEGINNING OF LOGIC

The logic of a computer can be traced back to a bone found from the European Ice Age, circa 32,000 B.C. from Blanchard, France, which is the earliest known human notation of marks on it. This stone has marks of what appears as finger and toe holes. The ancient man achieved simple counting with the utilization of fingers, toes, sticks, stones and marks in the sand or on cave walls to calculate the number of faces in the tribe, or the number of cattle in the herd and yes!!! even the number of wives owned.

Around 300 B.C the first mechanical calculating devises came into existence with the simple logic of sliding balls, which is known as the Chinese abacus (or suan pan), and the Japanese called it the soraban.

It is important to distinguish the early abacuses (or abaci) known as counting boards from the "modern" abaci. The counting board is a piece of wood, stone or metal with carved grooves or painted lines between which beads, pebbles or metal

Discs were moved. The abacus is a device, usually of wood (plastic, in recent times), having a frame that holds rods with freely sliding beads mounted on them.

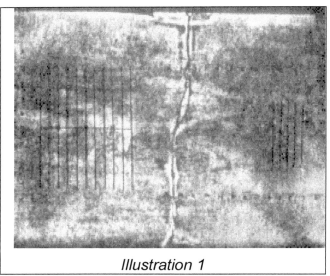

Illustration 1

The image in Illustration 1 is a picture of the oldest surviving counting board named the Salamis tablet, used by the Babylonians circa 300 B.C. It was discovered in 1899 on the island of Salamis. It is a slab of marble marked with 2 sets of eleven vertical lines (10 columns), a blank space between them, a horizontal line crossing each set of lines and Greek symbols along the top and bottom. A large crack runs down the middle; two sets of vertical bars on either side of the crack

ANCIENT TIMES

During Greek and Roman times, counting boards that survived are constructed from stone and metal similar to the sketch Illustration 2.

(The Roman Empire fell c. 500 A.D.)

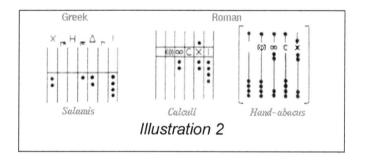

Illustration 2

MIDDLE AGES

Wood was the primary material from which counting boards were manufactured; the orientation of the beads switched from vertical to horizontal. As arithmetic (counting using written numbers) gained popularity in the latter part of the middle Ages, the use of the abacus began to diminish in Europe

Illustration 3

MODERN TIMES

The abacus as we know it today, appeared 1200 A.D. in China; in Chinese, it is called *suan-pan*. Beginning in about 1600 A.D., use and evolution of the Chinese abacus was begun by the Japanese via Korea. In Japanese, the abacus is called *soroban*. It is thought that early Christians brought the abacus to the East (note the vertical direction of both the *suan-pan* and the Roman hand-abacus from Illustrations 2 and 4).

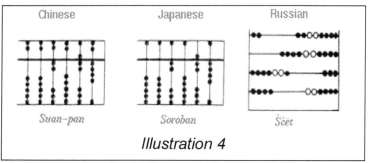

Illustration 4

Recent archeological excavations have revealed

a Mesoamerican (Aztec) abacus

(*Nepohualtzitzin*), circa 900-1000 A.D., where the

counters were made from kernels of maize threaded through strings mounted on a wooden frame.

Variations on a Theme

The *Illustration 5* image is a cover of a manual

published in 1958 by Lee Kai-chen, the inventor

of this "new" abacus designed with 4 decks

(essentially, it consists of 2 stacked abacii; the

Illustration 5

*

top abacus is a small 1/4 soroban and the bottom one is a 2/5 suan-pan). Lee Kai-chen claimed that multiplication and division are easier using this modified abacus and includes instructions for determining square roots and cubic roots of numbers.

The never-ending improvements of the counting techniques continued as time passed. The challenges of improving the mechanical methods of calculating, comparing and even entertainment continued thru history and finally the generation of the computer, as we know it in today's time has surfaced.

COMPUTERS ON THE RISE

On July 1 1890, two thousand clerks began processing the 1890 U.S. census, assisted by engineer Herman Hollerith's mechanized tabulating system. This event -- the most extensive information-processing effort ever undertaken -- launches the creation of the office-machine and paves the way for the founding of IBM (International Business Machines).

Illustration 6

Pictured in illustration 6 is a man operating a Holleriths Tabulating machine, which was used to calculate the 1890 Census. The Hollerith Electric Tabulating System is used again in the 1900 census, however, by 1905, Hollerith's interests have gravitated toward commercial ventures. He tailors his machines for processing information in business settings, and by 1911, he ran a prosperous punched-card office-machine company. That year, his health failing, Hollerith sells the business to a holding company, which in 1924 becomes the enormously successful IBM.

During the first half of the 1900's, numerous mechanical calculators were invented, built, and put into production. The demand for mechanical processing grew at a very speedy rate.

In 1950 a British mathematician and computer pioneer Alan Turing declared that one day there would be a machine that could duplicate human intelligence in every way and prove it by passing a specialized test. In this test, a computer and a human hidden from view would be asked random identical questions. If the computer were successful, the questioner would be unable to distinguish the machine from the person by the answers.

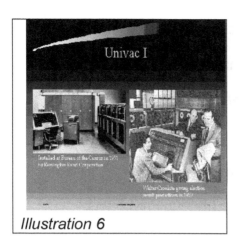

Illustration 6

Pictured on Illustration 6 is the first commercially available electronic computer named as the UNIVAC I. This unit surfaced in 1951 and was the first general-purpose computer designed to handle both numeric and textural information. These large computers occupied large rooms and had to be temperature controlled because the unit was sensitive to heat.

The idea of a computer to large volume

calculations and handle large volume text such

as letter writing became a desire for businesses,

but the cost for the large UNIVAC I was not

affordable to the smaller business. The need to

build a smaller computer was realized.

In 1953, it was estimated that there were 100

computers in the world.

Illustration 7

In 1976 the Apple pc was born.   The first

personal computer was housed in a plastic case

and include color graphics, the Apple II (pictured

in Illustration 7) was an impressive machine.

Also, with the introduction in early '78 of the

Apple Disk II, the most inexpensive, and easy to

use floppy drive ever (at the time). Today the

Apple computer is known as the Macintosh

computer.

Video games exploded in the 80's with action

games such as Space Invaders and a chewing

character called Packman. The game craze

continued for many years to come. Turning the

computer on and watching the little images scurry

across the screen hypnotically drew the attention

of the viewer for a challenge. One would spend

hours at a time with the challenge of defeating

the computer in the games. When defeating the computer, the user would acquire of sense of championship over such a complex machine in a way where the individual would feel something like a genius. Psychologists across the country expressed concerns that our children and adults alike were becoming addicted to video games. The visionary aspects of the games were little images of space creatures, talking buttons (referencing to pack man) and things of that nature. The logic was quite simple. It consisted of conditional logics of "If, And, Take and Give". Another way of putting it is subtract or add based on conditions.

During the 90's the race magnified with the Computer technology reaching more homes at a rapid rate as each year in the 90's passed. The nineties seemed like a computer epidemic. The convenience of the laptop was becoming into the scene. The craze of competition among friends, families, and business associates comparing who had the fastest computer, most memory, or largest hard drives. In 1995, a 50 MHZ speed processor was considered fast and a 1Gig hard drive was considered huge, and of course, not to mention adding 16mgs ram would be a screaming computer. Approx 5 yrs later when the Y2K approached, 700MHZ, 512 ram, and 20 gig hard drive would be considered the screaming computer. Many companies used this type of pc for a server. The year 2000 started with talk of a 1 gig processor in the industry. By the year 2000, we have reached the point where it seemed everyone had a pc computer in their home.

THE RACE IN THE COMPUTER INDUSTRY

The technology extended to the peripherals such as color printers, scanners, digital camera's, cd rom writer and high-speed modems, DSL, Broadband, flash drives etc.

The software industry growth struggled to keep up with the hardware industry, improving as years went by. The race was on to capture the general public's business by creating extremely user-friendly computers and software as well as affordable not only for the businesses but for home use also.

The pressure to be the first and best in the business was the cause of software being released with glitches causing computer failures. In order to correct the glitches, the software giants such as Microsoft made available what they called patches, and later referred to as service packs. Ultimately, this solution to software fixes has exploded across the software industry as a way to correct the glitches.

The computer industry finally successfully created user-friendly computers and software. In turn, the greed to seize the public 's business, the industry supplied their customers with free support that was not limited to simple tasks, nor did the industry take into consideration on what this unlimited support would result in.

Unlimited telephone and document support made available what seemed to be simple instructions on hardware installs to technical and non-technical individuals alike. The availability to the novice the combination of free tech support service, complex computer parts such as ram, hard drives, internal modems, and complicated utility programs such as Norton's and PC Tools had created a flood of self made computer professionals. Most of the self made computer professionals knew enough to be a certified tech's nightmare. As am example, I had a friend who bought a pc from a company that guaranteed her computer. She leaned by trial and error and calling support on how to service her computer. She proceeded on with her knowledge and started jumping on her friends computers as if she knew what she was doing and going into settings and changing things. Then when the computer would die, she'd get on the phone and call me.   Needless to say that was short lived, because after a while, I refused to enable her to think she knew all there was to know about computers when in fact, her actions actually harmed other peoples computers. The presumptuous idea to apprehend the marketplace in order to rank number one for the product, twisted into a financial nightmare for the industry.

What was intended to be an attempt to provide user-friendly systems, the industry provided a colony of individuals with a false sense of confidence that they can be a computer specialist without the need to take classes and know what actually makes the unit operate.   Even children as young as nine years old were dissecting pc's and installing electronic components as their parents proudly watched what appeared to be their gifted child because of his ability to tear apart this powerful electronic unit.

Naturally, the self-made computer technician with their proud self-confidence felt that they had mastered the computer and knew all there is to know about the once intimidating machine resorted to servicing other computers on a professional level without knowing the computer and electronic logic that is necessary to truly service a computer.   Thus, many times the self-made technician would attempt to repair the issue at hand instead of the cause, which would only result in a temporary fix, causing problems that are more complicated later on and even at times permanently damaging hardware.

As time passed, the computer industries realized a wake up call of the impact on self made computer techs. The repeated calls for support only created a financial burden on the computer industries. The industry soon realized that many times the support call was from a self made technician who caused a situation during a repair because of the lack of computer and electronic logic knowledge.   Many times the result would be extensive support help and often resulted with the vendor replacing hardware at no cost due to the self-made tech improperly configuring the item.

The expense of excessive use of tech support and the replacing of hardware at no charge was difficult for the industry to keep up, and therefore the industry had to resort to offer various support programs for a fee as a means to offset the industries losses.

Even with fees being charged for support, the industry realized that the cost of tech support for their products did not lessen the financial burden significantly because of the repeated calls from self-made techs and the returns of damaged hardware. With the need to get more professionals in the market and in the work place, the industry expanded the certification programs into specialty areas in order to isolate the professional from the self made technician.   This is where A+ Certification comes into play. This certification demonstrates the competency of a legitimate computer technician.

FINANCIAL IMPACT FROM THE PC COMPUTER RACE

The computer industry expanded at a very rapid rate. Computer sales extended to the local discount stores, which reduced the price of computers to very affordable prices. Not only did the discount stores offer better then affordable computers, they also offered excellent financing, such as interest free, or no payments for a year and credit card acceptances.

The smaller operating computer shops that once built custom computers felt the impact of the discount store sales of computers because they were unable to sell computers as inexpensive as the large discount stores and were not able to offer the many financing plans as the stores offered.   The large department stores were able to sell computer at lower prices because space allowed them to stock larger quantities of computers, which in turn the store paid less for the products with quantity pricing.

For a while there was trend on Internet Access Subcription sales where a free or very low cost computer was offered in exchange to committing to online services for a specific time.

As a result of the sales competition and the self-made techs working out of their basement with the availability of the hardware to the general public, numerous computer shops consequently fell out of business.

The computer evolution is responsible for the communications gap closing much tighter. Communicating on the web is far less expensive and makes us readily available to our family, friends and business associates.

This movement financially affected what was thought to be the giants of communications such as the telephone companies who felt the impact of the craze through loss of long distance business. Since the use of email and paying bills online became common practice in our every day lives, the Postal service was also affected with the decline of postage business. Both the telephone companies and the Postal service are striving to regain business loss by getting involved with the Internet Era.

Department stores have experienced and continues to realize the financial impact of computers with the Internet shopping capabilities, hence the main departments stores have also joined the march into world of the web by providing web shopping.

Newspapers were not exempt from loss of business to the Internet as the cyber space provided up to the minute news and weather on the home PC computer.

During the approaching of the historical year 2000 the scare of Y2k compliancy issues with the computer systems came into play. Corporations and government institutions across the country spent millions of dollars updating their systems and hiring computer specialist. It was the fear of Y2K that caused these corporations and government institutions to reach out to us professionals for help.

Due to the unnecessary excessive fear of computer failures in critical areas, citizens nationwide hoarded water, food, gas and candles. In the second half of the year 1999 electric generators were hard to find because fear of losing electricity from Y2K. Of course, the computer industry boomed temporarily as a result of the Y2K fear instead of the desire for progress.

The financial impacts mentioned above are just a mere sample of what transpired during the crossover to the millennium year. This could be a subject all of its own. When a company suffers financial loss, their employees and vendors also feel the blow.

Proudly I can say as a computer analyst/programmer from the 70's and a computer specialist in the millennium, as I predicted before the chime of the midnight bell, Y2K arrived, ALL IS WELL and we head off to continue the awesomeness world of electronic growth!

Bibliography

William R. Corliss. *Computers*.
U.S. Atomic Energy Commission,
Office of Information Services
Catalog #73-600074, 1973

The Moschovitis Group
History of the Internet, Moschovitis Website 1999

Louis Fernandes.
THE ABACUS, THE ART OF CALCULATING WITH BEADS
The Website of Louis Fernandes 1999

Lee Kai-Chen,
How to Learn Lee's Abacus, 1958

Glossary

Abacus: A binary device (like two five-fingered hands) using five beads on one side of wooden divider and two on the other side to count groups of five.

Binary:.

1.Characterized by or consisting of two parts or components; twofold.

2. Of or relating to a system of numeration having 2 as its base.

Computer.
A physical unit that has various electronic components within it to produce logical results

Hardware.

The associated physical equipment directly involved in the performance of data processing or communications functions

Logic.

1. The study of the principles of reasoning, especially of the structure of propositions as distinguished from their content and of method and validity in deductive reasoning.

2.a. A system of reasoning.

b. A mode of reasoning.

c. The formal, guiding principles of a discipline, school, or science.

3. Valid reasoning.

Software.
 The programs verbage, routines, and symbolic languages that control the functioning of the hardware and direct its operation utilizing logic.

www.ingramcontent.com/pod-product-compliance
Lightning Source LLC
LaVergne TN
LVHW052316060326
832902LV00021B/3929